Looking at Animal Parts

Let's Look at Animal Wings

by Wendy Perkins

Consulting Editor: Gail Saunders-Smith, PhD

Consultant: Suzanne B. McLaren, Collections Manager
Section of Mammals, Carnegie Museum of Natural History
Edward O'Neil Research Center, Pittsburgh, Pennsylvania

Mankato, Minnesota

Pebble Plus is published by Capstone Press,
151 Good Counsel Drive, P.O. Box 669, Mankato, Minnesota 56002.
www.capstonepress.com

Copyright © 2007 by Capstone Press. All rights reserved.
No part of this publication may be reproduced in whole or in part, or stored in a retrieval system, or transmitted in any form or by any means, electronic, mechanical, photocopying, recording, or otherwise, without written permission of the publisher. For information regarding permission, write to Capstone Press, 151 Good Counsel Drive, P.O. Box 669, Dept. R, Mankato, Minnesota 56002.
Printed in the United States of America

1 2 3 4 5 6 12 11 10 09 08 07

Library of Congress Cataloging-in-Publication Data
Perkins, Wendy, 1957–
   Let's look at animal wings / by Wendy Perkins.
   p. cm. —(Pebble plus. Looking at animal parts)
   Summary: "Simple text and photographs present a variety of animal wings and their uses"—Provided by publisher.
   Includes bibliographical references and index.
   ISBN-13: 978-0-7368-6716-0 (hardcover)
   ISBN-10: 0-7368-6716-3 (hardcover)
   1. Wings—Juvenile literature. I. Title. II. Series.
QL950.8.P47 2007
571.3'1—dc22   2006020937

Editorial Credits
Sarah L. Schuette, editor; Kia Adams, set designer; Renée Doyle, book designer; Charlene Deyle, photo
   researcher; Scott Thoms, photo editor

Photo Credits
Corbis/Charles & Josette Lenars, 18–19; Sea World of California, 16–17
Getty Images Inc./The Image Bank/Andy Rouse, cover
NHPA/Henry Ausloos, 15
Shutterstock/Kirsten Bauman, 8–9; Kiyoshi Takahase Segundo, 11
SuperStock/age fotostock, 12–13; James Urbach, 1, 20–21
Tom & Pat Leeson, 4–5
Visuals Unlimited/Gary Carter, 7

Note to Parents and Teachers

The Looking at Animal Parts set supports national science standards related to life science. This book describes and illustrates animal wings. The images support early readers in understanding the text. The repetition of words and phrases helps early readers learn new words. This book also introduces early readers to subject-specific vocabulary words, which are defined in the Glossary section. Early readers may need assistance to read some words and to use the Table of Contents, Glossary, Read More, Internet Sites, and Index sections of the book.

Table of Contents

Wings at Work 4
Kinds of Wings 10
Awesome Animal Wings 20

Glossary 22
Read More 23
Internet Sites 23
Index 24

Wings at Work

Animals use their wings
to fly and swim.
Animals send messages
to each other
with their wings.

A blue jay flies over a garden. It spots a butterfly sitting on a flower.

The butterfly's colorful wings give the blue jay a message. The bird knows this butterfly will not taste good.

Kinds of Wings

Hummingbirds flap their tiny wings very quickly.
These birds can fly forward and backward.

Condors fly with two wide wings. These birds soar on currents of air.

Puffins are birds that swim
with their wings.
They glide under the water.

Manta rays aren't birds,
but they have fins like wings.
These fish slowly flap their
fins to swim.

Some katydids have wings that look like leaves. These insects can easily hide from predators.

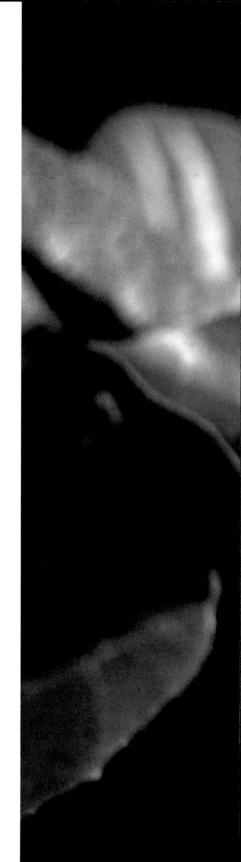

Awesome Animal Wings

Short or long,

feathered or smooth,

animals use their wings

in many ways.

Glossary

bird—a warm-blooded animal with two legs, wings, feathers, and a beak; most birds can fly.

current—a stream of moving air; condors ride on currents as they fly.

flap—to move up and down

glide—to move smoothly and easily

insect—a small animal with six legs and three body sections; most insects have wings.

predator—an animal that hunts another animal for food

soar—to fly very high in the sky

Read More

Dahl, Michael. *Do Whales Have Wings?: A Book About Animal Bodies.* Animals All Around. Minneapolis: Picture Window Books, 2003.

Miles, Elizabeth. *Wings, Fins, and Flippers.* Animal Parts. Chicago: Heinemann Library, 2003.

Ring, Susan. *Show Us Your Wings.* Bloomington, Minn.: Yellow Umbrella Books, 2004.

Internet Sites

FactHound offers a safe, fun way to find Internet sites related to this book. All of the sites on FactHound have been researched by our staff.

Here's how:

1. Visit *www.facthound.com*

2. Choose your grade level.

3. Type in this book ID **0736867163** for age-appropriate sites. You may also browse subjects by clicking on letters, or by clicking on pictures and words.

4. Click on the **Fetch It** button.

FactHound will fetch the best sites for you!

Index

blue jays, 6, 8
butterflies, 6, 8
condors, 12
fish, 16
flying, 4, 6, 10, 12
hiding, 18
hummingbirds, 10

insects, 18
katydids, 18
manta rays, 16
messages, 4, 8
predators, 18
puffins, 14
swimming, 4, 14, 16

Word Count: 137
Grade: 1
Early-Intervention Level: 15